W9-AQU-417

Animals in Order

Sharon Sharth

Sea Jellies

From Corals to Jellyfish

Franklin Watts - A Division of Scholastic Inc.
New York • Toronto • London • Auckland • Sydney
Mexico City • New Delhi • Hong Kong
Danbury, Connecticut

Dedicated to Sally Gallagher Schlaerth,
my mother

With special thanks to Kirk Fitzhugh, Ph.D., associate curator of Polychaetes Invertebrate Zoology at the Los Angeles County Museum of Natural History

Photographs ©: Animals Animals: 21 (K.Atkinson/OSF), 22, 23 (G.I. Bernard/OSF); BBC Natural History Unit: 41 top (Jurgen Freund), 25 (Constantinos Petrinos); Dembinsky Photo Assoc.: 15, 41 bottom, 43; Peter Arnold Inc.: 19, 27 (Kelvin Aitken), 5 top right, 13, 29 (Fred Bavendam), 1 (Kevin Schafer), 17 (Still Pictures); Photo Researchers, NY: cover, 31 (Charles V. Angelo), 7, 35 (A.Flowers & L.Newman), 42 (David R.Frazier/Photolibrary), 33 (David Hall), 37 (Andrew J. Martinez), 40 (Michael McCoy), 5 top left (NHPA/A.N.T.), 5 bottom left (Gregory Ochocki), 5 bottom right (Kjell B.Sandved/S.I. Washington, D.C.), 6 (Sinclair Stammers/Science Photo Library); Ulrich Technau: 39 (Darmstadt University of Technology).

Illustrations by Pedro Julio Gonzalez and A. Natacha Pimentel C.

The photo on the cover shows blade fire coral. The photo on the title page shows sea nettles.

Library of Congress Cataloging-in-Publication Data

Sharth, Sharon.
Sea jellies: from corals to jellyfish / Sharon Sharth; [Pedro Julio Gonzalez and A. Natacha Pimentel C., illustrators].
p. cm. — (Animals in order)
Includes bibliographical references and index.
ISBN 0-531-11867-3
1. Cnidaria—Juvenile literature. [1. Jellyfishes.] I. Gonzalez, Pedro Julio, ill. II. Pimentel C., A. Natacha, ill. III. Title. IV. Series.
QL375.6.S52 2001
592.5—dc21 2001017958

Printed in the United States of America.
1 2 3 4 5 6 7 8 9 10 R 11 10 09 08 07 06 05 04 03 02

Contents

OPEN OCEAN SEA JELLIES

ROCKY SHORE SEA JELLIES

CORAL REEF SEA JELLIES

SWAMP SEA JELLIES

What's a Sea Jelly?

The ocean is teeming with simple creatures that have no brain, gills, or backbone. Arms called *tentacles* loaded with stinging cells spring from their clear jellied bodies. Some of these sea jellies are as long as a 20-story building is tall! Others are so small you can barely see them. Some look like parachutes. Some look like balloons, fountain pens, or flowers. Although you might think that many sea jellies are plants, they are animals. The sea jellies in this book are *predators*, animals that hunt and eat other animals.

Look at the four sea animals shown on the next page. Only three are sea jellies. Can you tell which one is *not* a sea jelly?

Portuguese man-of-war

Sea pen

Brain coral

Squid

Traits of Sea Jellies

Did you pick the squid? You were right! How can you tell it is not a sea jelly? A sea jelly is a much simpler animal than a squid. Its soft, saclike body has a single opening through which food is taken in and waste matter is released. This opening is surrounded by tentacles. A sea jelly has no blood flowing through its body, nor does it have eyes, a heart, or other complex organs.

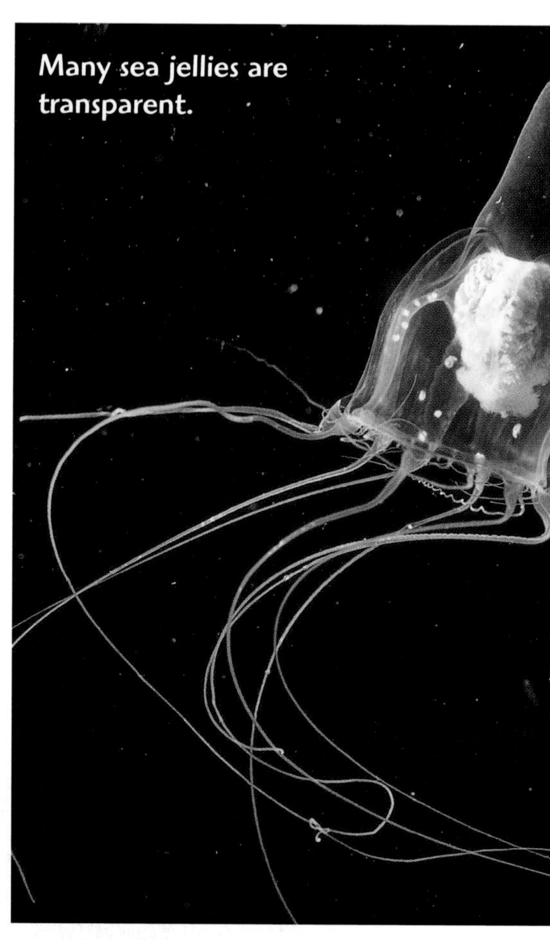
Many sea jellies are transparent.

Sea jellies have been drifting through Earth's oceans for 650 million years. They were thriving before dinosaurs existed! Many sea jellies are almost transparent, or see-through. Their bodies are jellied and flexible, not shaped by a skeleton the way a human's is.

There are sea jellies that live alone and those that live in groups called *colonies*. A sea jelly may take the shape of a tubelike *polyp* or of a bell-shaped *medusa*. Some may pass through both phases at different stages of their lives. Medusas are free-swimming. They drift with the water currents. Some can draw their bell together, and as they push the water out from under the bell, they pulse forward.

Polyps attach themselves to a surface, and many do not move. However, some can separate themselves from the rocks they cling to for support. They can "catch a current" to find a better feeding location or to avoid becoming another animal's meal. Others creep along in the sand. Some sea jellies swim by beating their tentacles or by shooting water out of their mouths. There are even some that can do somersaults to move short distances!

Some polyps can create a skeleton-like shell on the outside of their bodies. They pull calcium from the warm ocean water and turn it into a hard substance that helps protect them from predators. The outer skeletons of thousands of tiny polyps make up a coral reef.

The most important trait of sea jellies is their ability to capture *prey* or to defend themselves with their stinging cells. Housed in the tentacles, these cells release a poisonous harpoon that stuns, paralyzes, kills, or entangles their prey. The intensity of the effect depends on the animal. Most sea jellies are harmless to humans, but some can release *venom* strong enough to kill a person!

Sea jellies, such as this sea anemone, capture prey in their tentacles.

The Order of Living Things

A tiger has more in common with a house cat than with a daisy. A true bug is more like a butterfly than a jellyfish. Scientists arrange living things into groups based on how they look and how they act. A tiger and a house cat belong to the same group, but a daisy belongs to a different group.

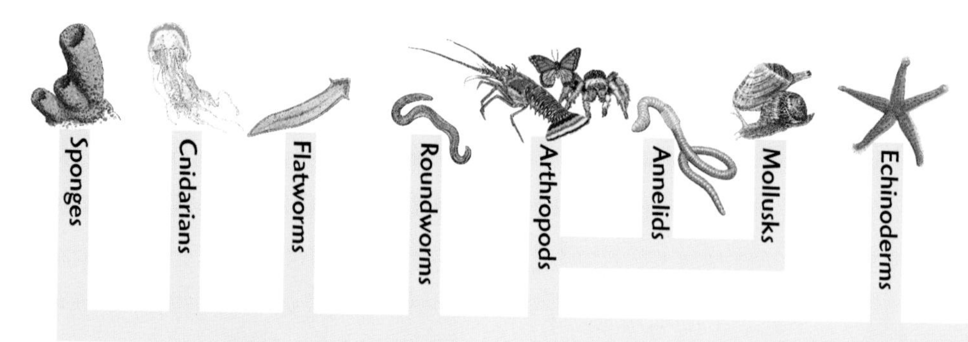

All living things belong to one of five groups called *kingdoms*: the plant kingdom, the animal kingdom, the fungus kingdom, the moneran kingdom, or the protist kingdom. You can probably name many of the creatures in the plant and animal kingdoms. The fungus kingdom includes mushrooms, yeasts, and molds. The moneran and protist kingdoms contain thousands of living things that are too small to see without a microscope.

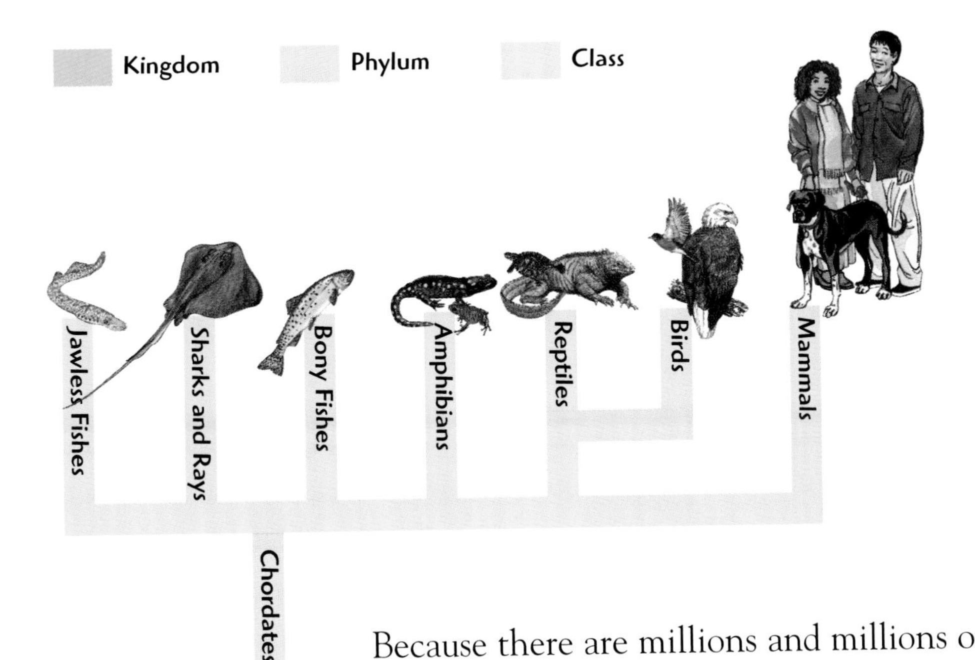

Because there are millions and millions of living things on Earth, some of the members of one kingdom may not seem all that similar. The animal kingdom includes creatures as different as tarantulas and trout, jellyfish and jaguars, salamanders and sparrows, elephants and earthworms.

To show that an elephant is more like a jaguar than an earthworm, scientists further separate the creatures in each kingdom into more specific groups. The animal kingdom is divided into nine *phyla*. Humans belong to the chordate phylum. All chordates have a backbone.

Each phylum can be subdivided into many *classes*. Humans, mice, and elephants all belong to the mammal class. Each class is divided into *orders*; orders are divided into *families*, families into *genera*, and genera into *species*. All the members of a species are very similar and can mate and produce healthy young.

How Sea Jellies Fit In

You can probably guess that sea jellies belong to the animal kingdom. They have much more in common with swordfish and snakes than with maple trees and morning glories.

Sea jellies belong to the cnidaria (ny-DAIR-ee-uh) phylum. The scientific name, cnidaria, comes from the Greek "sea nettle," which refers to their stinging cells. Cnidarians (ny-DAIR-ee-uhnz) are *invertebrates*, which means they have no backbone. Can you think of other invertebrates? Examples include snails, squid, lobsters, sea stars, sponges, and crabs.

The cnidarian phylum can be divided into three classes. The scyphozoans (sy-fuh-ZO-uhnz) include most of the bell-shaped jellyfish. Anthozoans (an-tho-ZO-uhnz) are jelly animals that look like flowers, such as corals and anemones. Hydrozoans (hy-druh-ZO-uhnz) form colonies that look like plants or feathers. Included in this group are hydroids and the Portuguese man-of-war. Scientists believe that there are more than 10,000 species of cnidarians, and they belong to a number of different orders. You will learn more about fourteen species of sea jellies in this book.

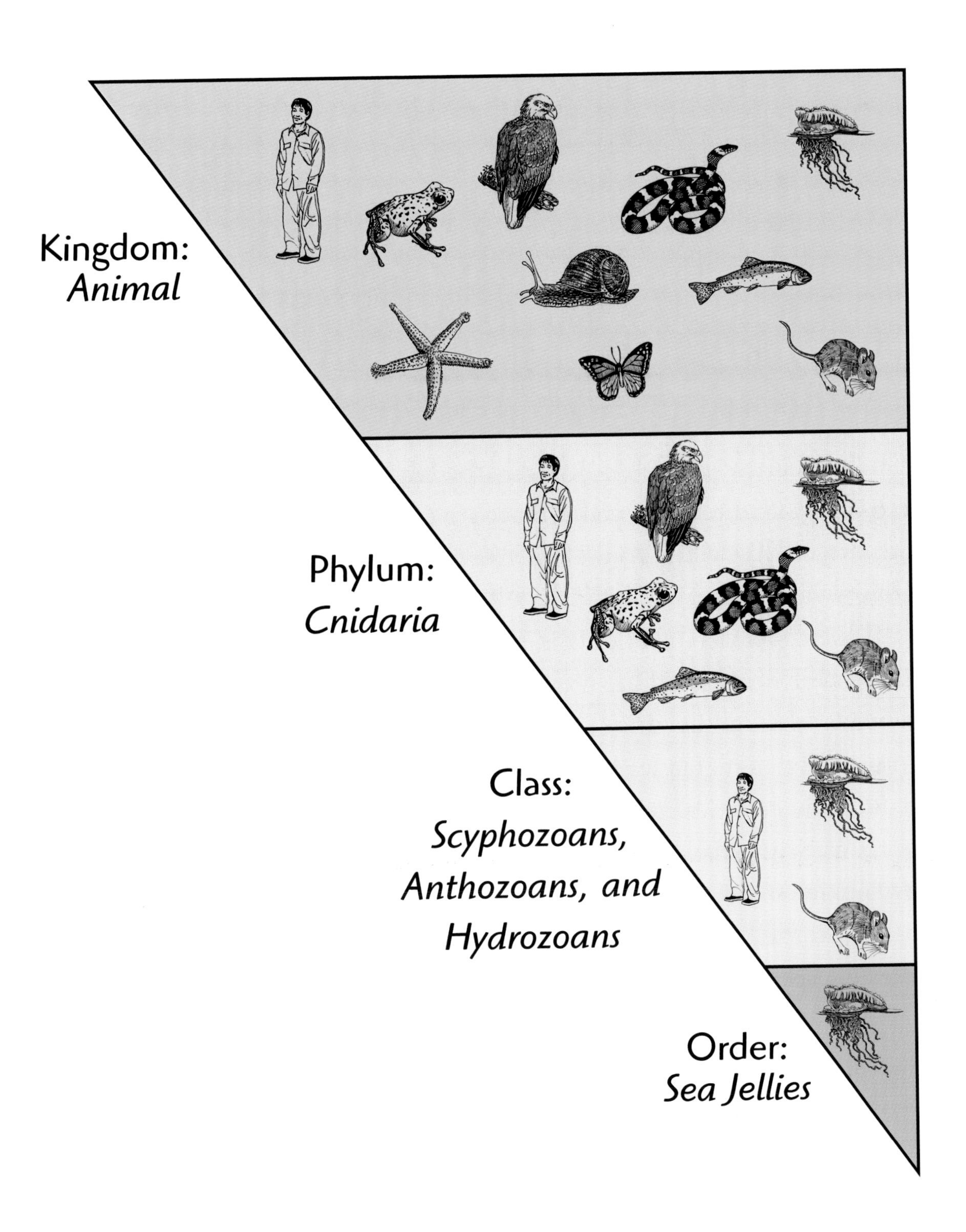
Kingdom:
Animal
Phylum:
Cnidaria
Class:
Scyphozoans,
Anthozoans, and
Hydrozoans
Order:
Sea Jellies

Jellyfish

FAMILY: Cyaneidae
COMMON EXAMPLE: Lion's mane jellyfish
GENUS AND SPECIES: *Cyanea capillata*
SIZE: 6.5 feet (2 m) across
Tentacles: 200 feet (61 m)

The lion's mane jellyfish is probably the biggest jelly, and it is one of the largest invertebrates on Earth. It is also the longest—its tentacles can extend 200 feet (61 m). This colorful reddish-brown giant can grow to be the size of a wading pool!

This sea jelly is found only in the cooler sections of the Atlantic and Pacific oceans and the North and Baltic seas. It doesn't have tentacles that fall like fringe around its bell, like many other jellyfish. Instead, the lion's mane jellyfish has 8 groups of tentacles under its dome. There are 150 tentacles in each group. This means that the lion's mane jellyfish has 1,200 tentacles! All those tentacles look like a windblown lion's mane.

When the lion's mane jellyfish hunts for prey, its huge, transparent bell slowly sinks. The net of tentacles spreads out around the jelly in the water. The tentacles release their barbs to capture and paralyze small fish, *plankton*, and anything else that it touches during its descent. When the lion's mane jellyfish has had its fill, it draws together its bell with a "whoosh" and rises. This poisonous sea jelly has been known to sting the uncovered skin of a diver.

Jellyfish

FAMILY: Ulmariidae

COMMON EXAMPLE: Moon jelly

GENUS AND SPECIES: *Aurelia aurita*

SIZE: 6 to 10 inches (15 to 25 cm) across

The moon jelly is a common form of jellyfish that lives in every ocean around the world. It has very short tentacles around its dome, and it is easy to spot because it has four horseshoe-shaped, purplish reproductive organs. The bodies of these jellyfish are transparent, so their organs show through the bell!

Young moon jellies can catch fish many times their size with their tentacles. As the jellies grow, they switch to feeding mainly on plankton. The plankton become stuck to the slimy liquid that covers the moon jelly's body. Tiny, hairlike particles move the plankton to the rim of the bell where there are eight special food pits. Then the jelly's arms scoop out the food and carry it to its mouth.

The female's eggs are fertilized by the male and develop in the frills of the tentacles near the female's mouth. As *larvae*, the eggs are set free into the ocean. They attach themselves to seaweed or rocks. There, they grow into small polyps with tentacles for catching food. In a process known as *budding*, growths form and separate from the polyp to become new

polyps. Each polyp creates a number of small larvae that eventually become adult moon jellies.

Moon jellies have one of the weakest stings of any jellyfish. It rarely will cause even a rash on an unsuspecting swimmer.

Jellyfish

FAMILY: Pelagiidae
COMMON EXAMPLE: Compass jellyfish
GENUS AND SPECIES: *Chrysaora hysoscella*
SIZE: 12 inches (30 cm)
Tentacles: 32 feet (10 m)

This transparent glassy-white jelly gets its name from the sixteen yellow or reddish-brown lines that radiate from the center of its flattened dome. It looks like the face of a compass. The compass jellyfish has twenty-four tentacles that hang from its dome while four thick and frilly longer tentacles dangle from the center of its body. It is found in seas where the climate is neither too hot nor too cold. It feeds on plankton and smaller jellies.

The compass jellyfish has both male and female reproductive organs. This sea jelly usually starts off as a male. Then it becomes male and female at the same time. During this phase, it can fertilize its own eggs. Ultimately, it becomes female.

Contact with this jelly's tentacles can lead to blisters and a high fever. Always be careful not to touch a beached sea jelly! A jellyfish may wash up on shore after a storm or when the tide is out. When this happens, it cannot get itself back to the water. Since a sea jelly is 95 percent water, it will quickly dry up in the sun. Don't get too close. Even when a sea jelly is dead, its stinging cells may still fire.

Portuguese Men-of-wars

FAMILY: Physalidae
COMMON EXAMPLE: Portuguese man-of-war
GENUS AND SPECIES: *Physalia physalis*
SIZE: 12 inches (30 cm)
Tentacles: 60 feet (18 m)

The Portuguese man-of-war drifts along the surface of the water. It looks like a lopsided purple balloon. Its long, coiled tentacles hang below the balloon, ready to shoot its barbs at any passing prey or at a hungry predator. The man-of-war's powerful sting is sometimes deadly. The poison released is similar to a cobra's venom.

Although the Portuguese man-of-war looks and acts as if it were a single animal, it is actually a colony of about 1,000 animals. Some members of the colony sting the man-of-war's prey. Others catch it. Some digest the food, and others lay eggs and produce new animals. Working together, they keep the colony strong, healthy, and running smoothly.

The man-of-war's balloon float is filled with gas, and it keeps the creature from sinking. The float also acts like a sail that helps the sea jelly move. The man-of-war keeps the float wet by tipping over on each side. This prevents the float from drying out in the hot sun.

This sea jelly was named after a Portuguese sailing ship.

Sea Pens

FAMILY: Pterioididae
COMMON EXAMPLE: Phosphorescent sea pen
GENUS AND SPECIES: *Pennatula phosphorea*
SIZE: 16 inches (40 cm) high

This fluffy sea pen looks like the ostrich plume George Washington would have used during his first year as president of the United States! Found in the Atlantic Ocean and the Mediterranean Sea, the phosphorescent sea pen anchors itself to the sandy or muddy bottom with its strong red stalk. The white polyps are attached to the feathered section. There are about 20 polyps in each row. Some kinds of sea pens may have up to 40,000 individual polyps!

The phosphorescent sea pen is known as a soft coral because its polyps cannot produce a hard skeleton the way that *stony corals* do. Soft corals are smaller and more delicate than stony corals. They also need less sunlight, so these sea jellies are found in deeper waters than stony corals.

A sea pen cannot swim but it can creep. It moves across the seabed as it slowly revolves on its stalk. The movement is so slow, it's almost impossible for a human being to see it.

The phosphorescent sea pen glows! Tiny grains are embedded in the slimy, sticky liquid that covers this coral. The liquid lights up a greenish blue when the sea pen is touched.

Anemones

FAMILY: Mesomyaria
COMMON EXAMPLE: Parasitic sea anemone
GENUS AND SPECIES: *Calliactis parasitica*
SIZE: 3 inches (8 cm) high

The parasitic sea anemone has 200 cream-colored tentacles circling its mouth opening. Its pale yellow body is streaked with lines of reddish-brown, and its thick jellied walls are shaped like a tube. This sea jelly can live on rocks or on snail shells. Sometimes the snail shell is inhabited by a hermit crab.

The hermit crab uses the stinging tentacles of parasitic anemones for protection from its enemies. The anemone uses the crab to move around and feeds on the crab's leftovers. A hermit crab can often be seen traveling with up to eight anemones covering its back. Hermit crabs have been known to try to eat the anemones they live with. Parasitic anemones have been known to try to eat the crabs too!

Some anemones use their strong tentacles to grab hold of prey such as fish, mollusks, and crustaceans. The anemone paralyzes its victim

and carries it to its tube to eat it. Some anemones even feed on dead organisms. When the lifeless particles touch the anemone, they become coated in the slimy, sticky liquid that covers the anemone's body. Then tiny, hairlike particles move the food to the anemone's mouth. Plankton is also eaten this way.

Fire Corals

FAMILY: Milleporidae
COMMON EXAMPLE: Fire coral
GENUS AND SPECIES: *Millepora tenera*
SIZE: 3 feet (1 m)

Found along the top and scattered around the sides of tropical reefs, fire coral is not a real coral. Like the Portuguese man-of-war, fire coral is actually a colony of sea jellies working together for the common good. Fire coral polyps come in two forms. Some polyps are feeders, and others are defenders. The feeding polyps have four tentacles around their mouth. Five or more defense polyps surround each feeding polyp. All polyps sink into tiny holes protected by the hard crust. They pop out quickly when food or danger appears.

Fire coral branches out in a tubelike web that develops from larvae. A chalky, limestone crust is secreted over rocks and dead corals. The top of the structure, near the surface of the water, does not become hard. Instead, it is covered with soft tissue that contains the stinging cells. Fire coral grows taller as the crust at its base thickens. The layers underneath die, and the polyps move up. A colony may grow to be 3 feet (1 m) high, so fire coral plays an important part in the development of the coral reef.

Fire coral gets its name from the blistering sting it inflicts on the human body. This mustard-colored cnidarian pierces the skin and causes itching and terrible pain.

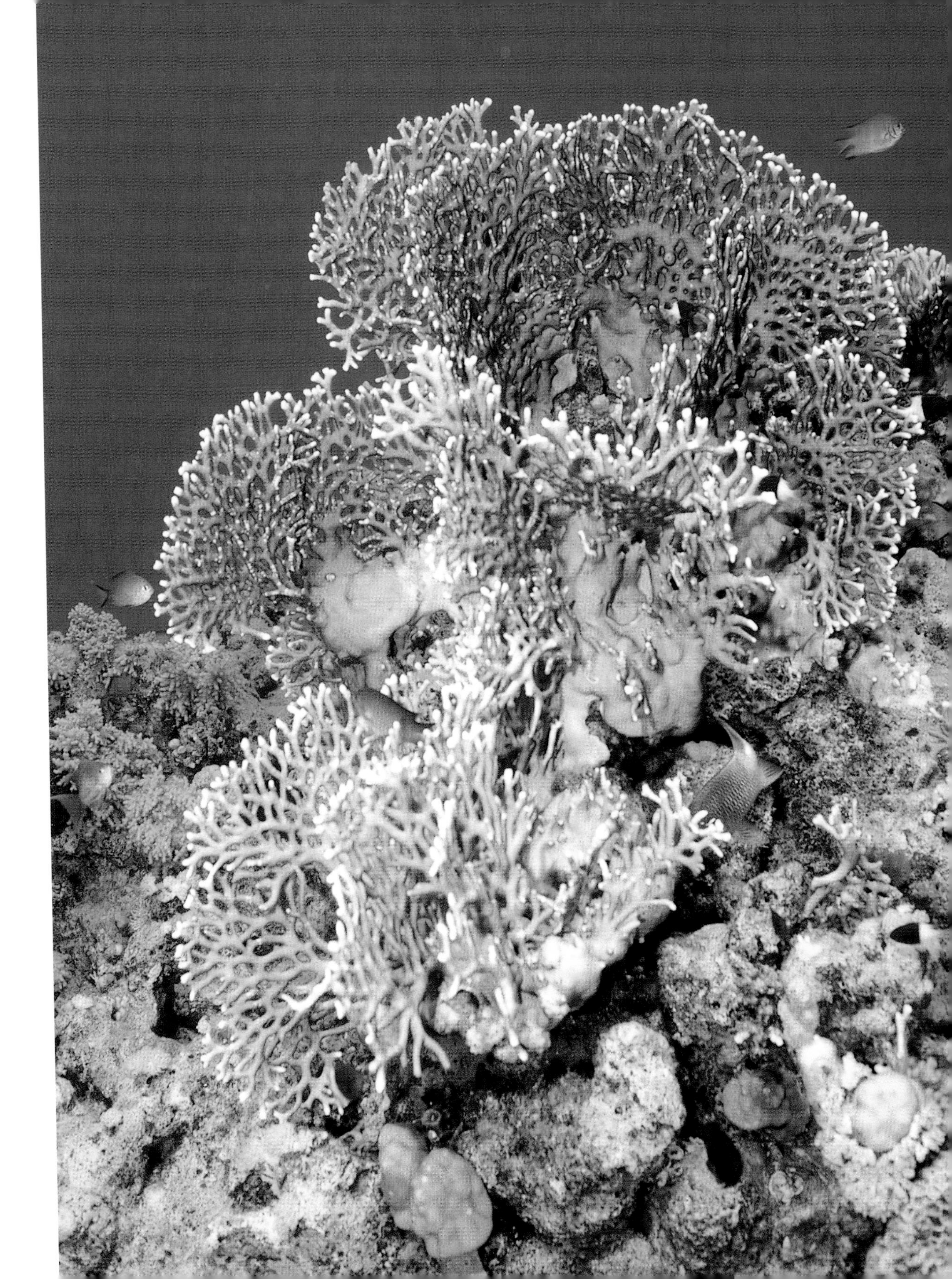

Sea Wasps

FAMILY: Chirodropidae
COMMON EXAMPLE: Sea wasp
GENUS AND SPECIES: *Chironex fleckeri*
SIZE: 6 inches (15 cm)
Tentacles: 10 feet (3 m)

Long tentacles hang from the sea wasp's translucent box-shaped body as it moves through the warm, tropical sea. It hunts for its favorite meal of shrimp along the bottom of shallow, muddy waters, so this creature may be hard to see. Watch out if you're swimming in waters near Australia, Japan, or the Philippine Islands.

The usual prey of a sea wasp is plankton and fish. It cannot attack humans, but it will sting anything that it meets. Sea wasps are more dangerous than sharks. The size of a coconut, a sea wasp can kill a human being within 3 minutes! Its venom comes from the stinging cells on its tentacles and is more deadly than any snake bite. When those tentacles—or even the bell—are touched, the stinging cells fire a harpoon of venom that causes severe burns. Its victim often will die in intense pain.

The sea wasp is the most venomous of all the sea jellies. It is also the fastest swimmer in the cnidaria phylum. Its cubed umbrella is able to draw together many times per second to jet the sea wasp forward. It can change direction quickly, and it can rush forward at speeds of more than 5 mph (8 kph) for short distances. That's fast for a sea jelly!

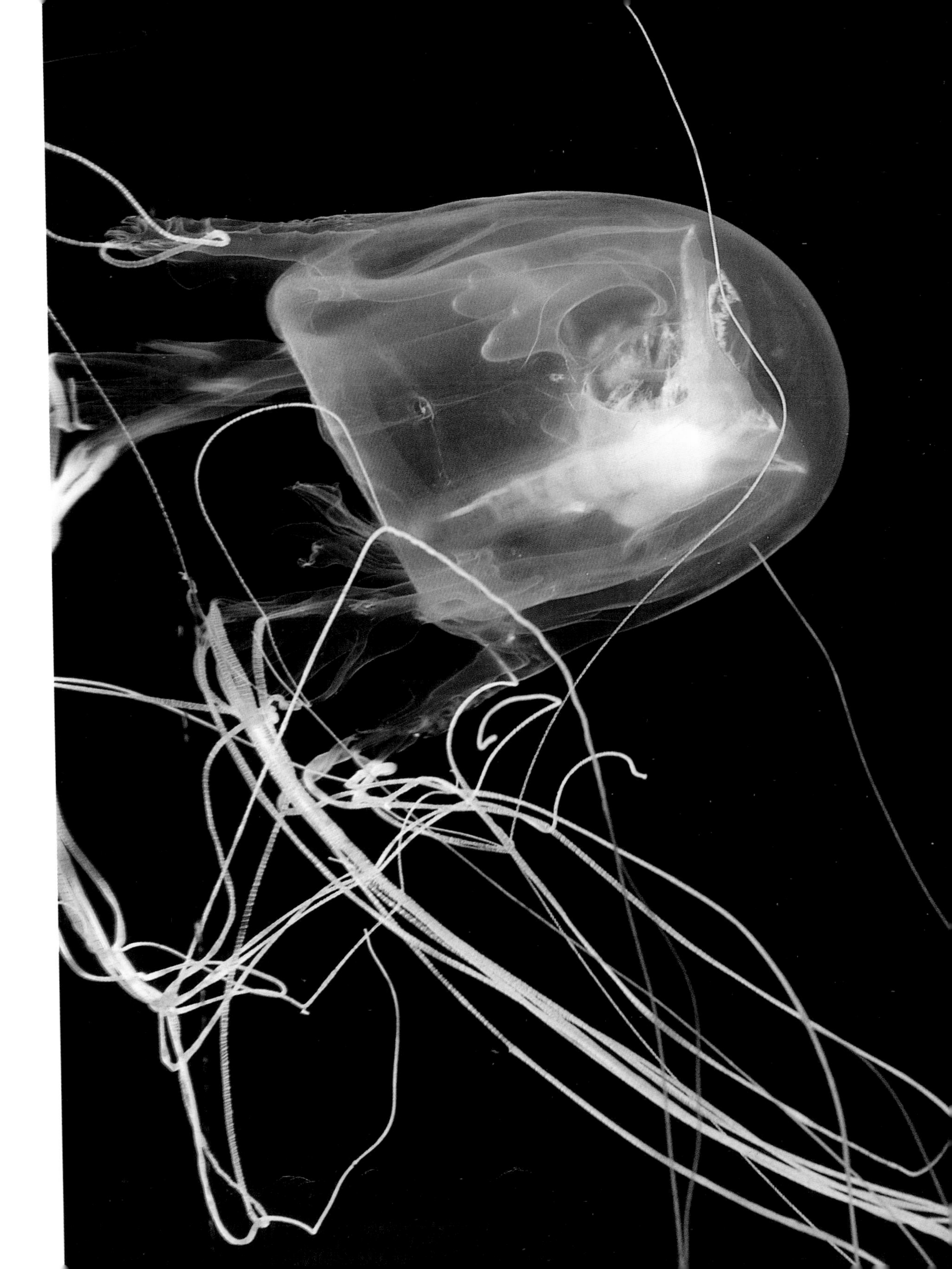

Anemones

FAMILY: Corymorphiidae
COMMON EXAMPLE: Ritter's sea anemone
GENUS AND SPECIES: *Heteractis magnifica*
SIZE: 3 feet (1 m)

Ritter's sea anemones add to the coral reef's vibrant beauty with their rust-colored bodies and green tentacles. These sea jellies are found in areas with strong water currents. They are one of the largest tropical anemones.

Anemones provide a haven for many other marine animals. Crabs and anemonefish, or clownfish, can live within the safety of the anemone's stinging tentacles without being harmed. As young fish, the anemonefish wiggles against the tentacles enough times to become immune to the anemone's poison. Then they can freely swim, feed, and reproduce while being protected by the anemone.

The anemone's sting keeps away the predators of the anemonefish. In return, the anemonefish attract prey for the anemone and protect it from small predators such as sea worms. Anemonefish also help to keep the sea jelly clean. They remove and eat parasites and dead tissue.

Ritter's sea anemones vary their diet depending on what is available to them. Plankton and fish are common meals for these animals.

Corals

FAMILY: Faviidae
COMMON EXAMPLE: Brain coral
GENUS AND SPECIES: *Diploria strigosa*
SIZE: 1 to 4 feet (30 cm to 1.2 m)

Brain coral looks like a brain! These huge greenish-gray or brown corals form a round colony as they wind their way over rocks and other hard surfaces.

Many sea jellies, including brain corals, reproduce and expand their colonies by budding. New polyps bud near the parent coral's mouth. The buds don't separate from the parent. More and more buds are added, and a long twisting line is formed with tentacles on each side.

The stinging cells on their tentacles are used only for feeding. Like most stony corals, these cells are too weak to harm humans or to protect the coral from its enemies.

During the day, the tiny polyps are not visible. When it turns dark, they pop out of their tiny holes to catch prey. They repeat the cycle with the morning light. Food, especially plankton, is more available at night. Corals feed on the plankton that moves from the depths up to the water's surface. The corals stretch their bodies to grab at the plankton and bring it toward their mouths. Like other sea jellies, corals eat other animals and digest them quickly.

Corals

FAMILY: Fungiidae
COMMON EXAMPLE: Mushroom coral
GENUS AND SPECIES: *Anthomastus ritteri*
SIZE: 6 inches (15 cm)

Mushroom corals are not the reef-builders that their relatives are. As soft corals, mushroom corals cannot build limestone skeletons for protection or strength. Instead, these circular sea jellies have flexible bodies that can change into different shapes! When approached by a predator, the mushroom coral can close itself down until it looks like a mushroom.

When the mushroom coral feeds, it looks very different. Its poisonous tentacles spread out in the water. As it grasps at the passing animal plankton, it looks like a flower or a pom-pom. Catsharks often lay their egg cases on these outstretched stalks of mushroom corals.

Mushroom corals live alone and do not form colonies the way the reef-builders do. They live in shallow, quiet waters and loosely attach themselves to the sandy ground. These sea jellies can slowly creep along the seabed. They can also lift off and float away if they are threatened or hungry.

Corals

FAMILY: Antipathidae
COMMON EXAMPLE: Black coral
GENUS AND SPECIES: *Antipathes dichotoma*
SIZE: 3/4 inch to 19.7 feet (2 cm to 6 m)

Black corals form colonies that look like plants or small trees. Many of the firm, black stalks attach their flat bases to rocks or other hard objects. Others may sink into the sandy ocean bed. Delicate branches shoot out from these stalks. Short, tubelike polyps cover them. Black corals may grow to be almost 20 feet (6 m) high!

Each black coral polyp has six tentacles that are always extended. These slow-growing corals send out their stinging cells to capture food, such as plankton. Black corals have been known to live for 20 years. They begin to reproduce at about 10 years old. They settle in caves, under overhangs, in deep water, and in other dark areas because they don't need sunlight to survive.

Black corals are important to people for many reasons. Some forms of black coral are believed to have medicinal or magical properties. For years, people have created jewelry with black corals. Earrings, bracelets, cuff links, and rings made with black coral are popular items. With a hammer or an ax, divers often will cut apart the coral and float the bounty up to the water's surface using air bags. The number of these precious sea jellies is declining as locals and foreign divers overharvest black corals from the reefs where they grow.

Jellyfish

FAMILY: Cassiopeidae
COMMON EXAMPLE: Cassiopeia jellyfish
GENUS AND SPECIES: *Cassiopeia xamachana*
SIZE: 5 to 7 inches (13 to 18 cm)

Cassiopeia jellyfish live in the warm mangrove swamps of tropical seas. Thousands of these sea jellies turn upside down to lie on the shallow, muddy bottom with their ruffled arms reaching upward. A small mouth is on each of these arms. In the still waters of the swamp, Cassiopeia jellyfish use their arms to capture and feed on plankton. They have no tentacles hanging from their bells as other jellyfish do. They don't need the tentacles' harpoons to catch prey. They get most of their nutrition from the *algae* that live in their jelly.

Cassiopeia jellyfish follow the sun. If this jellyfish is without sunlight for too long, it will shrink. It needs the sun to keep its partner, the algae, healthy and strong. In exchange for protection and sea jelly wastes, the algae give the Cassiopeia oxygen, food, and their greenish tint.

These sea jellies avoid being swept away by the currents with the help of an area on their bell that is curved inward like the inside of a bowl. It produces a suction that keeps the Cassiopeia in place. Cassiopeia jellyfish will stay in the same place unless disturbed by rough waters or predators.

Anemones

FAMILY: Edwardsiidae

COMMON EXAMPLE: Starlet sea anemone

GENUS AND SPECIES: *Nematostella vectensis*

SIZE: 3/4 to 2 1/2 inches (2 cm to 6 cm) high

The starlet sea anemone lives alone. It is found only in salty marshes in the northern hemisphere. This long, thin sea jelly cannot burrow in the sand. Instead, it burrows in the fine mud of the marsh and often buries its entire body. Sticking out of the mud are the sixteen to twenty white-spotted tentacles ready to inflict their venom on any passing prey.

The starlet sea anemone has a greyish-white coloring. Other colors are added depending on what it has eaten recently. Its last meal can be seen right through its jelly! Plankton and fly larvae are two of the starlet sea anemone's favorite prey, but this sea jelly will eat whatever its tentacles can bring down.

Since it has no known predators, the starlet sea jelly may sometimes be seen fastened to sea grasses or lying completely exposed on the mud. It can withstand a high salt content in the water and a wide range of temperatures. It can even be revived after being frozen in ice for several hours!

Saving the Coral Reefs

This coral reef is in the Solomon Islands, near Australia.

Coral reefs are the largest structures ever built by living organisms. The surprising thing is that they have been erected by tiny coral polyps! Coral polyps can extract calcium from the ocean water and turn it into a hard, limestone skeleton. Millions of these simple creatures grow on top of the remains of earlier colonies. Huge coral reefs have been formed in this way over thousands of years.

Coral reefs are one of the most densely populated areas on Earth. These "rain forests of the sea" provide a *habitat* for thousands of animals and supply the food they need to stay alive. Reefs protect nearby islands and mainlands from heavy surf and flooding. They offer economic stability to local communities through the fishing and tourist trades. New medicines for bone grafts and treatments for leukemia and skin cancer have been discovered in the coral reefs. Snorkelers and scuba divers flock to the reefs to observe their vibrant beauty close-up.

A diver explores a coral reef.

But coral reefs are in trouble. They are being destroyed at an alarming rate. Coral reefs grow only a fraction of an inch each year. Still, erosion by waves during hurricanes and other storms, along with rapid drops in temperature, injure the reefs. Delicate polyps cannot survive in these circumstances. Even a rise in the water's salt level can be deadly to some polyps.

Yellow, tubelike sponges are invading this coral reef.

Predators also ruin large sections of this delicate environment. The parrot fish crushes the hard outer shells of the corals with its powerful beak. The exposed, soft polyps become the parrot fish's feast. Sponges use strong chemicals and invade. They wipe out entire coral colonies to make room for themselves. Sea stars, especially crown-of-thorn sea stars, gather in huge swarms and crawl over the colonies. These animals have stomachs that come right out of their bodies! They throw their stomachs over the polyps and eat them. They leave only the dead, white skeleton of a once thriving community. Sea snails, worms, crabs, shrimps, many kinds of fish, and even some barnacles feed on the coral polyps.

However, people are the coral reef's worst enemy. Corals are damaged by boats, anchors, and careless divers. Global warming contributes to warmer seas and causes algae to leave their polyp homes. The polyps then may die of starvation. The abundance of carbon dioxide, pollution, overfishing, overcollecting by the jewelry industry,

Drilling for oil affects coral reefs.

underwater dynamiting, oil-drilling—all affect reproduction, feeding behavior, and growth rates of the corals.

But steps are being taken to preserve the world's coral reefs. Scientists have joined with architects to develop artificial reefs in hopes of learning how to stimulate new reef growth. Parks and reserves have been established. Guidelines to monitor and manage the reefs have been set up by conservation groups. In most countries, the exportation of corals is restricted or against the law. However, it is difficult to enforce these laws. Coral reefs play an important role in the health and stability of Earth. It is up to us to protect and preserve these amazing animals.

When marine biologists study coral reefs, they learn how to protect them.

Words to Know

alga (plural **algae**)—any of a large group of simple plants that are mainly aquatic and usually contain chlorophyll but do not produce seeds and lack roots, stems, and leaves

budding—a form of reproduction in which a bud grows from its parent and breaks off when it has grown into a fully formed organism

class—a group of creatures within a phylum that shares certain characteristics

colony (plural **colonies**)—a group of animals that lives together and works as a single unit

family—a group of creatures within an order that shares certain characteristics

genus (plural **genera**)—a group of creatures within a family that shares certain characteristics

habitat—the environment where a plant or animal lives and grows

invertebrate—an animal that does not have a backbone

kingdom—one of the five divisions into which all living things are placed: the animal kingdom, the plant kingdom, the fungus kingdom, the moneran kingdom, and the protist kingdom

larva (plural **larvae**)—the first stage in the life cycle of some kinds of animals

medusa—a cnidarian with a bell-shaped dome

order—a group of creatures within a class that shares certain characteristics

phylum (plural **phyla**)—a group of creatures within a kingdom that shares certain characteristics

plankton—a group of tiny floating plants and animals

polyp—a tubelike animal with tentacles that surround its mouth opening

predator—an animal that hunts and eats other animals

prey—an animal that is hunted and eaten by other animals

species—a group of creatures within a genus that shares certain characteristics. Members of the same species can mate and produce young.

stony coral—a coral that secretes a hard skeleton around itself using calcium from the water

tentacles—long, flexible, stinging body parts that circle the mouths of cnidarians

venom—a poison that animals use to catch prey or to fight enemies

Learning More

Books

Cerullo, Mary M. *Coral Reef: A City that Never Sleeps*. New York: Cobblehill, 1996.

Landau, Elaine. *Jellyfish*. Danbury, CT: Children's Press, 1999.

Owens, Caleb. *Coral Reefs*. Chanhassen, MN: Child's World, 1998.

Schaefer, Lola M. *Sea Anemones*. Mankato, MN: Pebble Books, 1999.

Wu, Norbert. *A City Under the Sea*. New York: Atheneum Books for Young Readers, 1996.

Web Sites

Oceanic Research Group

www.oceanicresearch.org/cnidarian.html

Brief descriptions about different sea jellies accompany clear, colorful photographs.

Sea World

www.seaworld.org/coral_reefs/introcr.html

This site provides information on all aspects of a coral's life. It covers its eating habits, reproduction, habitats, causes of death, and conservation of coral populations.

University of California, Berkeley

www.ucmp.berkeley.edu/cnidaria/cnidaria.html

Featuring an in-depth introduction to the world of cnidarians, this site focuses on ecology, history, and fossils. It includes interesting facts about these animals and great photographs.

Index

About the Author

Sharon Sharth is a certified scuba diver and has worked with dolphins at the Kewalo Basin Marine Mammal Laboratory in Hawaii. She has written many books about animals for children including *Jellyfish*, *Squid*, *Hawks*, *Robins*, *Finches*, *Rabbits*, and *Whale: A Sticker Safari*. She has written a book about navigation, *Way to Go: Finding Your Way with a Compass*, and *The Book of Telling Time*. Ms. Sharth has been involved with children for many years as a teacher and as a Screen Actors Guild *Book Pal* reader. She believes that, through our children, the creatures with whom we share the Earth will finally come to be protected and respected.